ELECTRICITY

SCIENCE & TECHNOLOGY

ELECTRICITY

ELECTRONICS

GADGETS & DEVICES

INTERNET

ROBOTICS

SPACE EXPLORATION

ELECTRICITY

Mason Crest

Mason Crest

Mason Crest
450 Parkway Drive, Suite D
Broomall, PA 19008
www.masoncrest.com

Series ISBN: 978-1-4222-4205-6
Hardback ISBN: 978-1-4222-4206-3
EBook ISBN: 978-1-4222-7599-3

First printing
1 3 5 7 9 8 6 4 2

Cover photograph by Gjp311/Dreamstime.com.

Library of Congress Cataloging-in-Publication Data
Names: Mason Crest Publishers, author. Title: Electricity / by Mason Crest. Other titles: Electricity (Mason Crest Publishers)
Description: Broomall, PA : Mason Crest, [2019] | Series: Science & technology
Identifiers: LCCN 2018034422| ISBN 9781422242063 (hardback) | ISBN 9781422242056 (series) | ISBN 9781422275993 (ebook)
Subjects: LCSH: Electrical engineering--Juvenile literature.
Classification: LCC TK148 .E37 2019 | DDC 621.3--dc23 LC record available at https://lccn.loc.gov/2018034422

QR Codes disclaimer:
You may gain access to certain third party content ("Third-Party Sites") by scanning and using the QR Codes that appear in this publication (the "QR Codes"). We do not operate or control in any respect any information, products, or services on such Third-Party Sites linked to by us via the QR Codes included in this publication, and we assume no responsibility for any materials you may access using the QR Codes. Your use of the QR Codes may be subject to terms, limitations, or restrictions set forth in the applicable terms of use or otherwise established by the owners of the Third-Party Sites. Our linking to such Third-Party Sites via the QR Codes does not imply an endorsement or sponsorship of such Third-Party Sites, or the information, products, or services offered on or through the Third-Party Sites, nor does it imply an endorsement or sponsorship of this publication by the owners of such Third-Party Sites.

CONTENTS

KEY ICONS TO LOOK FOR

Words to Understand: These words with their easy-to-understand definitions will increase the reader's understanding of the text, while building vocabulary skills.

Sidebars: This boxed material within the main text allows readers to build knowledge, gain insights, explore possibilities, and broaden their perspectives by weaving together additional information to provide realistic and holistic perspectives.

Educational Videos: Readers can view videos by scanning our QR codes, providing them with additional educational content to supplement the text. Examples include news coverage, moments in history, speeches, iconic moments, and much more!

Text-Dependent Questions: These questions send the reader back to the text for more careful attention to the evidence presented here.

Research Projects: Readers are pointed toward areas of further inquiry connected to each chapter. Suggestions are provided for projects that encourage deeper research and analysis.

Series Glossary of Key Terms: This back-of-the-book glossary contains terminology used throughout this series. Words found here increase the reader's ability to read and comprehend higher-level books and articles in this field.

WORDS TO UNDERSTAND

alloy a metal that is made by combining one or more metals

alternative something that can be used in place of something else

capacitor a device that stores electricity

cautious to be careful

chemical reaction a reaction that causes change when chemicals combine and form different substances

constantan an alloy of copper and nickel

contraction a movement of muscles when they become tighter

deplete to become less

electric shock a painful feeling that a person experiences when electricity suddenly passes through the body

emission the act of sending out gases or heat into the atmosphere

fatal dangerous to life; something that can cause death

fluorescent bulb a bulb that produces light when an electric current is passed through it

fluctuation to change frequently

fossil fuel a fuel that is produced by the burning of fossilized objects like coal, which is made from dead trees and plant matter

friction a force that is produced when one object rubs against another object

geyser a machine that is used in many households for heating water

global warming increase in the temperature of the earth because of the rise in greenhouse gases like carbon dioxide

gravitational force a force that makes two objects with mass move toward each other

incandescent bulb a bulb that produces light as a result of being made very hot

manganin an alloy of copper, manganese, and nickel

mechanical energy an energy that is produced by an object due to its motion or its position

maglev train a high-speed train that runs with the help of magnetic force

natural gas a gas that is found inside earth and is used for cooking and heating

neon a colorless and odorless gas that shines brightly when electricity flows through it

nuclear energy an energy that is stored in the nucleus of atoms; also known as atomic energy

nuclear reactor a machine used for producing nuclear energy in the form of electricity

pollution the contamination of air, water, or soil by means of harmful or poisonous chemical substances in the environment

porcelain a ceramic substance used for making crockery

radioactive element an element that contains very harmful energy that is usually produced during nuclear reactions

renewable something that can replace itself by a natural process

replenish to enrich

resistor a substance that regulates the flow of electric current in an electric circuit by providing resistance

smokestack a large chimney through which gases, smoke, and vapors discharge

synthetic cloth a type of cloth that is made of synthetic fibers like polyester or nylon

thermostat a device used in machines or appliances to control or regulate temperature

turbine a machine that contains a wheel that is made to rotate by the extreme pressure of any fast flowing liquid or gas and which produces power

welding to heat and hammer metals to give them a desired shape or size

Electric Charge

An electric charge is a basic property of matter. Matter is made up of atoms as its smallest unit. An atom is further divided into electrons, protons, and neutrons. A proton has a positive charge while an electron has a negative charge. Thus, an electric charge is produced when electrons and protons move through an atom.

INTRODUCTION

Electricity is a form of energy that is created by the flow of electric charges. The word electricity was coined by a physician William Gilbert. Electricity is neither a renewable nor nonrenewable source of energy. The concept of electricity is also observed in several creations of nature. Lightning is the most common natural example. Another example, a fish called the electric eel, has a high voltage current in its cells to catch its prey. In addition to all these examples, all the activities in animal and plant cells take place due to electric stimulation inside the nerve cell. Electricity has several applications in today's world. It is used to run appliances, such as televisions, computers, microwave ovens, washing machines, and more.

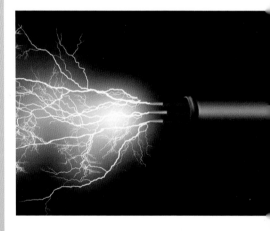

Law of conservation of charge

All the electric charges follow a universal law. This is the *law of conservation of charges*. This law states that the total charge in an isolated system is always constant. Thus, just like energy, charges can neither be created nor destroyed. However, the charges can move and be transferred from one part of the object to another.

Properties of electric charges

Electric charges have several unique properties:

- Like charges repel each other, while unlike charges attract.
- An electric charge is denoted by the symbol e. It has a value equal to 1.6×10^{19}.
- Motion of an object does not affect its charge.
- Electric charges are always conserved.
- The total amount of charge on an object is the sum total of all the charges spread in the different directions of the object.

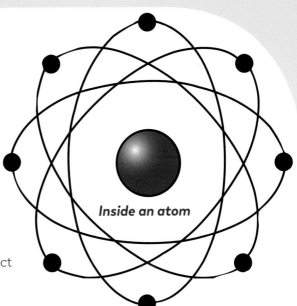

Inside an atom

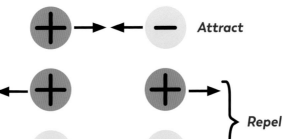

Attract

Repel

SCIENCE FACTS

- **Contraction** of heart muscles takes place due to the presence of electric current in the body.

- The neutron of an atom is neutral. It does not have any kind of charge on it.

Static Electricity

Static electricity simply refers to the electricity produced by static charges because of **friction** between two objects. Static charges are stationary or immovable charges. These charges collect on the upper surface of an object and remain there unless disturbed. An object acquires a static charge when it is rubbed against another object.

Benjamin Franklin

Lightning Rod

Benjamin Franklin invented the lightning rod in 1752. A lightning rod is a metal rod that is placed at the top of a building. The other end of the rod is connected to the ground with a wire. Lightning rods protect buildings from the effects of lightning.

SCIENCE FACTS

- A bolt of lightning can measure up to three million volts (3,000 kv)!

- **Synthetic clothes** develop more static charges than cotton fabrics.

Lightning

Lightning is the discharge of electricity in the atmosphere. It is an example of static electricity. The particles present inside the cloud rub against each other and get oppositely charged. These clouds are then attracted by other clouds or by the surface of the earth. This causes lightning.

Applications

Static electricity is widely used in **pollution** control plants and **smokestacks**. The dust particles are given a charge and collected on the filter. It is also used in photocopy machines and while painting cars. The ink and spray that have the charged particles are pasted or copied on the material.

Electric Current

An electric current is defined as the amount of charge flowing through a wire at a given time. This form of flowing energy is known as current electricity. There is just one difference between static electricity and current electricity. Static electricity occurs due to the collection of charges whereas current electricity exists due to the flow of charges. Electricity is measured in amperes and it is denoted by the symbol A.

Heating Effect of Electric Current

When an electric current is passed through a conductor, there is a resistance in the flow of electric current and this generates heat. This heating property of electric current is used in appliances like electric immersion water heaters, electric irons, etc. All of these have a heating element in them, which is generally made of specific **alloys** like nichrome, **manganin**, **constantan**, etc. A good heating element has high resistivity and high melting point. An electric fuse is an example of the application of the heating effect of an electric current.

Converting heat to electricity

SCIENCE FACTS

- Alternating current was used for the first time by a French neurologist, Guillaume Duchenne in 1855.

- Electricity travels at the speed of about 186 miles (300 km) per second, similar to light.

Types of Electric Current

Depending on the flow of charges, electric current is further divided into two types. These are alternating current and direct current. Alternating current, or AC, is the current which changes its flow direction after every fixed time span. On the other hand, direct current, or DC, is the type of current which does not change its flow direction.

High voltage transmission lines deliver power from electric generation plants over long distances using alternating current.

Sources of Electricity

Electricity is the most commonly used form of energy. It is used to light and warm homes, and to run all kinds of machines and appliances. Most of the electricity that we use is obtained by burning coal, petroleum, and **natural gas**.

Nonrenewable Sources of Energy

Nonrenewable sources of energy are sources that are available in nature in a limited amount. These sources are also known as non-**replenish**able sources of energy and need to be used carefully. They include **fossil fuels**, such as coal, oil, and gasoline. Even **nuclear energy** is another nonrenewable source of energy. It is produced by the **chemical reaction**s of **radioactive element**s in a nuclear power plant. It produces a large amount of energy that is used to produce electricity.

Renewable Sources of Energy

Renewable sources of energy are sources that are freely available in nature and can be used again and again. These sources are also known as replenishable sources of energy. These sources are solar energy, wind energy, water energy, biomass, and tidal energy. These sources of energy can be used repeatedly without depleting them.

Why Use Renewable Sources

Using renewable resources of energy is very beneficial for the environment as it poses minimum threat. Fossil fuels like coal, oil, natural gas, and nuclear fuels used to make electricity cause a lot of air pollution. They even contribute to **global warming** to a large extent. However, we can use renewable resources repeatedly without depleting them, and they do not contribute to global warming as there is no polluting **emission**, and they even have low-cost applications.

Conductors and Insulators

There are different types of materials that behave differently when electricity is passed through them. Some of the objects allow electricity to pass through them, while some do not. Based on this, there are different types of substances known as conductors, insulators, semiconductors, and superconductors.

Conductors

Conductors are substances that allow electricity to pass through them. Conductors have free, movable electrons that can easily carry electricity. Substances like aluminum, platinum, gold, silver, and water are good conductors of electricity. They are widely used in wires, coils, and electrical appliances.

Semiconductors

Semiconductors are materials that have conductivity values between conductors and insulators. They can be easily converted to an insulator or a conductor according to the need. Silicon, selenium, and germanium are the components that make a semiconductor. They are widely used as a part of the circuit.

Superconductors

A superconductor is an element, compound, or intermetallic alloy that will conduct electricity without resistance when cooled below a certain temperature. Superconductors can allow electrical current to flow without any energy loss and the type of current produced is called a supercurrent.

SCIENCE FACTS

- There are two types of semiconductors. P-type semiconductors are positively charged, and n-type semiconductors are negatively charged.

- A diode is the simplest type of semiconductor, while a transistor is the most complex kind of semiconductor.

Insulators

Insulators are substances that do not allow electricity to pass through them. Insulators do not have free, movable electrons to carry electricity. Instead their electrons are tightly packed. Substances like glass, **porcelain**, plastic, and rubber are some commonly used insulators. They are often used as a safety measure to protect against electricity.

Magnetism

Magnetism is a property displayed by magnetic substances wherein the substance can attract other substances with magnetic properties. A magnet can be a piece of metal, ore, or stone. They may occur naturally or can be made. Lodestone is a naturally occurring magnet.

History of Magnets

Magnets come in a variety of shapes. Some of the most common shapes include a bar, horseshoe, and ring. People have known about magnets for thousands of years. The word *magnet* comes from the ancient Greek town Magnesia. The ancient Greeks and Chinese were perhaps the first to know about the great attraction power of magnets. Historical records mention the use of magnets by the Chinese as far back as 2500 BCE.

SCIENCE FACTS

- Magnets are widely used in medical equipment, water heaters, computer disks, etc.

- When a magnet is cut into numerous pieces, each piece has a north pole and a south pole.

Poles of a Magnet

A magnet has two poles known as the magnetic poles. These are the north and the south poles. Magnetic poles always exist in pairs. Magnetic fields travel from one pole to the other. If two similar poles are kept together, the magnets repel each other. However, if opposite poles are kept together, the magnets attract each other.

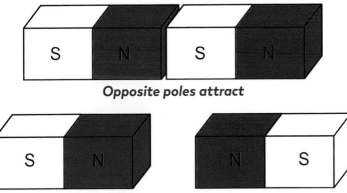

Opposite poles attract

Similar poles repel

Magnetic Field

When the electrical charges move through an object they produce magnetic fields. A magnetic field is an empty space created around a moving charge. This empty space has the capacity to attract other objects. A magnetic field travels in circles. It cannot be seen with the naked eye but its presence can be felt.

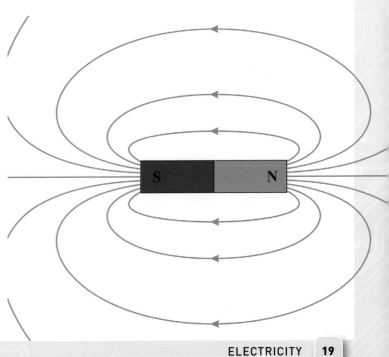

Electromagnetism

Electromagnetism is a phenomenon that occurs in electromagnets. Electromagnets are substances that behave as magnets when electricity is passed through them. Thus, both electricity and magnetism are important for an electromagnet.

A Fundamental Force

Electromagnetic force is one of the four fundamental forces. Fundamental forces are the natural forces that explain the interaction between objects. The other fundamental forces are strong force, weak force, and **gravitational force**. Any other force that exists in nature is derived from one of these forces.

Formation

When an electric current is passed through a wire, it produces a magnetic field. If this wire is wrapped around an iron substance, then the substance also becomes magnetic. This produces an electromagnet. If the number of coils of the wire is increased, then the magnetic property of the substance can be increased.

- An electromagnet is a temporary magnet and exhibits magnetism only when electricity is passed through it.

- Unlike magnetism, there are ways to switch electro-magnetic substances on and off.

Applications

There are several uses of electromagnets in present-day life. They are a key component of motors, speakers, electromagnetic cranes, **maglev trains**, and MRI equipment. They are also used in magnetic separators in research and medicine. Electromagnets are used in scrap separation too. Large circular electromagnets are used to separate iron.

Transformers

Transformers are objects that help in transforming electrical energy. The transformation of electricity can take place from higher voltage to lower voltage or from lower voltage to higher voltage. Voltage is the difference between electrical charges at any two points of a circuit.

Working of a Transformer

A transformer is made up of two coils. The primary coils are linked to the switch and obtain the incoming AC. This electricity generates a magnetic field in the coil. This magnetic field in turn produces an electric current in the secondary coil. This secondary coil is connected to the equipment.

Uses of Transformers

Transformers are widely used with heavy electrical appliances. The electricity that is distributed by power plants is not uniform. Electrical appliances can be harmed by such varying voltages. Thus, transformers help to produce electricity of constant voltage.

SCIENCE FACTS

- A transformer can only be used for alternating current (AC). It cannot be used for direct current (DC).

- If a transformer has a fewer number of turns in the secondary coil then it is a step-down transformer; however, if it has a greater number of turns then it is a step-up transformer.

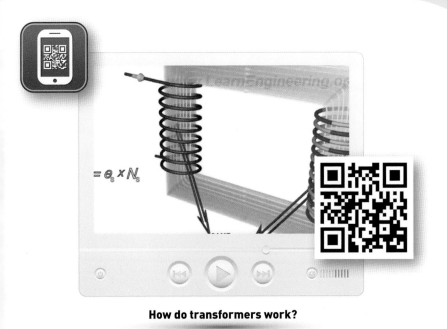

How do transformers work?

Types of Transformers

Depending on the voltage of electricity needed, there are two types of transformers. These are step-up and step-down transformers. Step-up transformers increase the voltage from lower to higher voltage. They are used in the production of x-rays. Step-down transformers help in reducing the voltage from a higher level to a lower level. These transformers are often used for **welding** purposes.

Electric Circuits

An electric circuit is the path through which electricity flows. Electricity flows only through a closed electric circuit. For instance, when a light is switched on, the bulb glows due to the flow of electricity from the switch to the bulb. This implies that when the light is switched on, the circuit is closed, thus flow of electricity takes place. However, when it is switched off, the circuit is opened and the flow of current breaks, and the light turns off.

Series Circuit

A series circuit is a type of electric circuit. It allows electricity to flow in one direction only. In this circuit, every component connected to the circuit receives the same amount of electricity. For instance, the bulbs shown here are switched on by a series circuit.

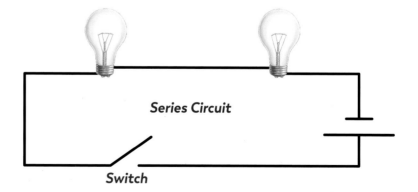

Series Circuit

Switch

Parallel Circuit

A parallel circuit allows the flow of electricity in more than one direction. Even if one of the paths is broken, the circuit works through some other pathway. Electronic gates are the most common example that operate through parallel circuits.

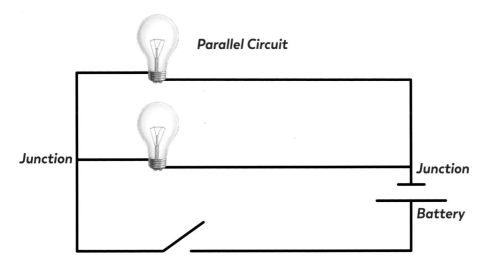

Parallel Circuit

Junction

Junction

Battery

Series-parallel Circuit

A series-parallel circuit is a mixture of both series and parallel circuits, and is used in industries. Some components of the circuit are connected in a series while the rest are connected in a parallel circuit. This type of circuit enables the system to work even if some of the components of the circuit becomes nonfunctional. In a series parallel circuit, more than one path exists for the current to flow.

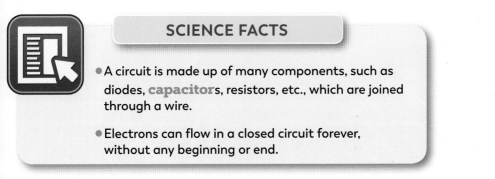

SCIENCE FACTS

- A circuit is made up of many components, such as diodes, **capacitor**s, resistors, etc., which are joined through a wire.

- Electrons can flow in a closed circuit forever, without any beginning or end.

Electric Motors

Electrical motors convert electrical energy into **mechanical energy**. They use electricity to perform mechanical work. Electrical motors are made up of innumerable coils. The rotation of these coils produces magnetic fields and currents. Thus, these rotations convert electrical energy into mechanical energy.

AC Motors

Electric motors that are driven by alternating current are known as AC motors. AC motors have one stationary part around which a rotating coil of wire is fixed. This coiled wire receives electricity for the motor. The amount of electricity supplied decides the speed of an AC motor. Appliances such as mixers and hand drills use AC motors.

Types of AC Motors

Broadly, AC motors are divided into two types: single-phase motors and three-phase motors. Three-phase motors are used mostly in factories, while single-phase motors are used in homes. Single-phase motors are more popular than three-phase motors.

DC Motors

Michael Faraday invented the DC motor in 1821. It runs on direct current electricity. DC motors can be divided into three types: brush motors, brushless motors, and stepper motors. Amongst the three, brush motors are the most common. DC motors have a wide variety of applications. Electric razors, remote control cars, and electric car windows are a few common appliances that use DC motors.

SCIENCE FACTS

- The speed of a motor can be altered by increasing the number of wire coils.

- Electric motors are used in electric cars.

Switches

Switches are electrical devices. They help in closing and opening an electric circuit. When a switch is turned on, it completes the circuit. This allows electricity to pass through. However, when a switch is closed, it breaks the circuit. This stops the flow of electricity.

Types and Parts

There are many types of switches. They are distinguished based on the way they operate. Toggle, push button, selector, and joystick are a few of the different types of switches. The process of operating a switch is known as actuation. The part of the switch that is pressed, or operated, is called the actuator. The actuator controls the opening and closing of a circuit.

Toggle Switches

A lever or a handle actuates toggle switches. Such switches work when their lever is either pushed or pulled. These switches are used where one has to control a large amount of electric current or mains voltages. The common light switch is the most common example of toggle switches.

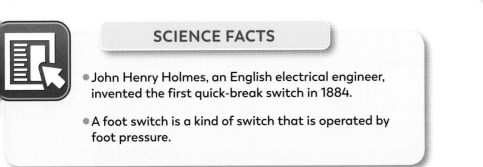
Biased Switches

Biased switches, or mechanical switches, have a push button to turn them on and off. The two types of biased switches are push-to-make and push-to-break switches. Push-to-make switches turn on when pressed, while push-to-break switches turn on when released.

Signal Switches

Signal switches are the switches that function only during certain situations. These switches are specially designed and respond during special times. Such switches are widely used in security systems, such as alarm systems and anti-theft systems where the switches respond only during abnormal situations.

Tools and Equipment

Electricity is the most common source of energy. People are well aware of its existence; however, they often do not know how to handle it carefully. Therefore, one should be **cautious** while handling electrical appliances and equipment. Also, one must know the tools and equipment used with electricity.

Wire Strippers

Wire strippers are metal tools used to cut the plastic or rubber coating of the wire. Wire strippers are used when there is a need to expose the naked wire. This may be required to make a connection with some other wire or electrical appliance. However, a naked wire should never be touched before turning off the switch.

SCIENCE FACTS

- An antistatic wristband is a protective tool that electricians wear on their wrist.

- Pliers are one of the basic tools used by electricians. They are used to cut, twist, and grip wires.

Voltage Detector

A voltage detector is used to check the flow of current in a wire. There are different types of voltage detectors. The simplest is a light bulb, or **neon** lamp and **resistor**, placed in series with the power supply or mains electricity. Other detectors include a screwdriver with a built-in LED, and plug-in devices with LEDs.

Screwdrivers

Screwdrivers are one of the most commonly known tools required for work related to electricity. They are used for tightening or loosening the screws attached to an electronic object. Screwdrivers are available in different shapes and sizes.

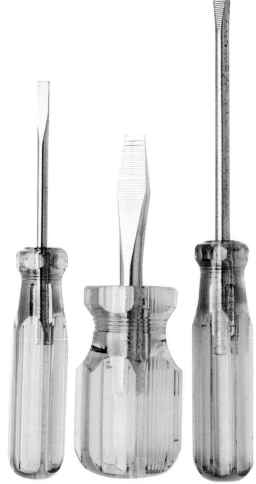

Common Electrical Appliances

A wide variety of electrical appliances make our day-to-day work easy. From heating water and cooking food to cleaning our homes we depend on these appliances. Electrical appliances are widely used in industries as well.

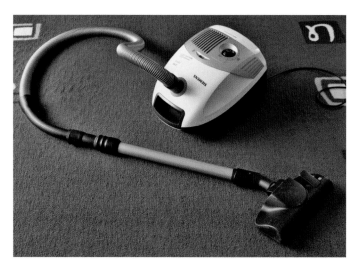

Vacuum Cleaners

Vacuum cleaners are household appliances. They are used for cleaning purposes and trapping the dust. The device is comprised of an electrical motor, a suction pump, and filter bags. The motor helps in conversion of energy. Vacuum cleaners attract the dust particles through suction mechanism.

SCIENCE FACTS

- A generator is an electrical appliance that converts other types of energy into electrical energy.

- The suction in vacuum cleaners is carried out by the current of air produced in them.

Food Processors

Food processors are another common household electrical appliance. It is used for several purposes in the kitchen. It helps in chopping, slicing, and grinding raw fruits, vegetables, and other food items. It also consists of an electric motor that converts electrical energy into mechanical energy. Food processors have blades and other small tools to process the food items. These tools work when they get an electric current.

Electrical Fans

An electric fan runs on an electric motor. The motor converts electric energy into rotational energy. This rotational energy further rotates the blades of the fan resulting in the cooling of the room. The blades of the fan and motor are attached to a hub.

Safety

In the present day world, electricity has various applications. It has proven to be truly beneficial for man. However, playing with electricity can also be dangerous. Therefore, it is essential to keep in mind safety rules while working with electricity or other electrical appliances.

Electric Shock

Depending upon the intensity of electricity, an **electric shock** can be minor or major. A minor shock can cause a sensation in the body; however, a major shock can prove to be **fatal**. In case of an electric shock, the source of electricity should be shut down immediately. The person should be touched with covered hands. During an electric shock, the body tends to get stuck to the source from which the shock is being produced. Touching with bare hands a person who is receiving the shock can transfer the shock. In cases of electric shock, a doctor should be consulted immediately.

SCIENCE FACTS

- In the United States, more than 4,000 people are killed or injured every year by electricity.

- While working with the service panel, the main power switch should always be turned off.

Effects of Current

The human body is a good conductor of electricity. Therefore, while using an electrical appliance, it is important to wear rubber footwear, rubber or plastic clothing, rubber gloves, and to carry a wood piece along. Also, any water supply should be cut off while working with electric **geysers** or water heaters. Since water is a good conductor of electricity, it may give an electrical shock.

Fires

Electrical fires are one of the most common accidents. Drawing an excess of electricity from a wall socket can result in fire. Therefore, electrical appliances should be kept away from rugs or beds, especially when they are switched on. Additionally, lit candles that are on top of electrical items can cause fire. In case of a fire, an extinguisher should be used immediately to extinguish the fire.

Measuring Instruments

Various devices can be used to measure electric charge, electric current, voltage, resistance, electrical conductance, capacitance, inductance, electric fields, and magnetic fields.

Voltmeter

A voltmeter is a measuring instrument that measures the voltage between any two points in a given electrical circuit. A voltmeter is constructed in such a way that it uses minimal electricity to perform its functions. The voltmeter measures voltage in volts (V). It can measure the voltage of both AC and DC.

Using a multimeter to measure electricity

SCIENCE FACTS

- The galvanometer was the first, basic type of ammeter.

- If used in homes, a wattmeter can be used to measure the total energy.

Ammeter

The flow of current in a given electrical circuit is measured using an ammeter. Like the voltmeter, an ammeter can measure the current both for AC and DC voltages. Ampere is the unit in which the current is measured using the ammeter.

Wattmeter

A wattmeter is an instrument that measures the power in a given electrical circuit. Power is defined as the rate of doing work. As the name suggests, the measurements made through a wattmeter are expressed in watts (W). However, a higher model of the instrument can also display the measurements like volts and amperes.

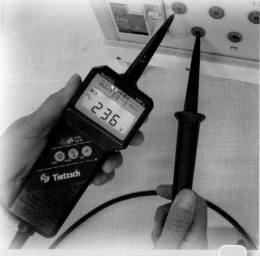

Consumption of Electricity

Consumption of electricity is the total amount of electricity that is used for any kind of work. Besides fuels like gasoline, electricity is another important source of energy that is used for a majority of work. Therefore, it is important to keep a check on the consumption of electricity and make efforts to reduce it.

Insulation

Insulating walls and ceiling helps keep a place warm. Therefore, it can reduce the dependence on the heater. The heater is one of the appliances that consumes a higher amount of electricity as compared to other appliances such as a fan. Therefore, insulating the walls and ceiling can reduce electricity consumption by 20 to 30 percent.

Setting the Thermostat

Setting the **thermostat** correctly in accordance to climate conditions can also help in reducing the consumption of electricity.

SCIENCE FACTS

- A compact **fluorescent bulb** uses only one quarter the electricity of an ordinary **incandescent bulb**.

- The energy that is wasted to light an office overnight is equivalent to the energy to heat water for one thousand cups of tea.

Maintaining Air Conditioners

The air conditioner is another appliance that consumes a large amount of electricity. If air conditioners are not properly maintained, they tend to have dirty filters. These filters further force these appliances to consume more electricity to give the desired results. Approximately 5 percent of energy could be saved by maintaining a clean air conditioner.

Alternative Electricity

With the increase in population, the global demand for electricity has increased greatly. At present, electricity is produced either by plants that burn fossil fuels or by nuclear power plants. However, both methods use nonrenewable sources of energy and cause pollution. Therefore, **alternative** means of energy should be used.

Solar Energy

Solar energy is a major source of energy. The energy obtained from sunlight is known as solar energy. In order to produce electricity using solar energy, the energy is collected through solar cells. Large panels of solar cells arranged in a grid form are known as solar panels. Electricity produced like this is pollution free, but is an expensive option.

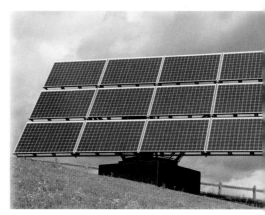

Wind Energy

Wind energy is another renewable source of energy. In this form of energy, blowing winds are used to rotate wind turbines, which results in the production of electricity. However, this form of energy cannot always be used, since the wind varies.

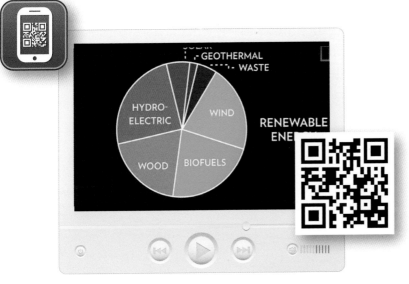

National Geographic: Alternative energy

Hydroelectricity

Electricity that is produced by flowing water is known as hydroelectricity. It is a renewable source of energy. Hydroelectricity is produced when a large amount of water hits rotating **turbines** with pressure. This results in the production of electricity that can be used for further tasks.

SCIENCE FACTS

- Around one fifth of the world's electrical demand could be met by water alone.

- Approximately 25 percent of the electricity required in San Francisco is produced by wind energy.

Conservation of Electricity

Generating electricity with the help of fossil fuels and **nuclear reactors** is an easy and quick process. However, both fossil fuels and radioactive elements are limited in nature. These materials may someday be **depleted**. Generating electricity using alternative energy sources is slow and expensive. This further strengthens the need to conserve electricity.

Remember to Unplug

Unplugging appliances when they are not in use can help in conserving electricity. Electricity can also be conserved if several plugs are replaced by a single power strip. Chargers that are not charging anything should also be removed from the outlet, as a connected charger continues to use energy.

Energy-efficient Appliances

Most modern electrical appliances come with energy efficiency ratings. It is better to buy products with higher ratings. Often these products may be a little expensive but they go a long way in conserving energy and thereby reduce the amount of money spent on energy expenses. For example, it is always better to use compact fluorescent bulbs (CFLs) instead of incandescent lamps, which consume a lot more energy.

Smart Ways

Appliances that are not being used should be turned off. For instance, a television should be turned off instead of being kept on standby. This is because televisions tend to consume electricity while on standby. Such measures can conserve a good amount of electricity. Similarly, room lights should be turned off when not being used. A thermostat is an electrical device that helps to maintain a constant flow of current without any **fluctuation**. Its temperature could be set according to the need. It should not be set at a very high or low temperature; rather, it should be set in accordance to the room temperature.

TEXT-DEPENDENT QUESTIONS

1. What did Benjamin Franklin have to do with electricity?

2. Name two good materials for conducting electricity.

3. What two ancient civilizations does the text say experimented with magnetism?

4. Name two types of circuits mentioned in the text.

5. What do switches open and close?

6. What are five types of electrical equipment?

7. What do wire strippers remove?

8. What does an ammeter measure?

9. What are three types of alternative renewable energy?

10. What are three types of nonrenewable energy?

RESEARCH PROJECTS

1. Read about Thomas Edison, Nikola Tesla, and George Westinghouse. Prepare a short report about their contributions to the study and spread of electricity in society.

2. Find out more about maglev trains. How do they work? How do they use electricity for power? Do they go faster than ordinary trains; if so, why? Write a report showing how maglev trains might have an impact on transportation in your state.

3. Read about how your community or city is using renewable energy. Has the amount it uses gone up? What sources does it use? Prepare a graph or chart showing how much of each type is used.

4. Find an electrical bill for your house or apartment. Read it over and find out what the symbols and information mean. How is your family charged for electricity? How can you reduce the electricity you use to pay less?

FIND OUT MORE

Books

Kallen, Stuart. *How Electricity Changed the World*. San Diego: Reference Point Press, 2018.

Morkes, Andrew. *Electrician (Careers in the Building Trade: A Growing Demand)*. Broomall, PA: Mason Crest, 2018.

Platt, Charles. *Make: Design Your Own Circuits: 17 Exciting Design Ideas for New Electronics Projects*. San Francisco: Maker Media, 2018.

On the Internet

History of Electricity
instituteforenergyresearch.org/history-electricity/

US Energy Use
www.eia.gov/energyexplained/index.php?page=electricity_user

Electricity Basics
www.edisontechcenter.org/basics.html

alloy a substance made up of a mixture of metals

capacitor a device that stores electricity

emission the act of sending out gases or heat into the atmosphere

digital expressed as series of the digits 0 and 1

friction a force that is produced when one object rubs against another object

hydraulic using powered created by water or liquid

interactive providing output based on input from the user

interplanetary a space mission that is planned for study of other planets

magnetic field a region around a magnet that has the ability to attract other magnets

microprocessor a very small circuit used in computers that performs all the functions of a central processing unit (CPU)

navigation the science of directing the course of a vehicle

programmable able to be given instructions to do a task

protocol a set of rules that is used by computers to communicate with each other across a network

renewable something that can replace itself by a natural process

rechargeable something that can be charged again and again

sensor a mechanical device that is sensitive to some signal and helps in responding to it

voltage difference in electric tension between two points

synchronize to operate two or more devices at the same time

viable capable of working successfully

INDEX

Photo Credits

Photographs sourced by Macaw Media, except for: Dreamstime.com: Daniel K. Malloy 11TR, Kobackpacko 29BR; Shutterstock: Mikhail Romanov 28BR.